How old is old?

A sense of time distinguishes Man from all other animals. Prehistoric and primitive man show by their burial customs a concern for the future and a respect for the past, and modern civilised man puts a great deal of effort into trying to reconstruct and understand bygone days.

The methods of the historian and archaeologist are familiar, so that, whether a reconstruction of Victorian Britain or of Neolithic Egypt is presented, it is not difficult for laymen to follow the lines of evidence. But the methods of the geologist working out the history of the Earth are a mystery to many people, and so they are not wholly convinced by his reconstructions. Geological events are so far back in time that the question is constantly asked: how can he be be sure? This booklet attempts to explain both the methods and the results of this geological detective work. It will show how the geological time scale has been built up and how the age of the Earth itself is being explored.

Hundreds and even thousands of years are periods of time readily understood by most people. Geologists use 'a million years' as a unit when dealing with Earth history, and talk casually of 50 or 500 million years, time-periods quite outside our comprehension. Figure 5 relates the time scales of the historian, archaeologist, and geologist to distances around the world. If one hundred years is equivalent to 40 metres, then the 5000 years of recorded history would extend 2 kilometres – twenty minutes' walk. Most archaeologists work within a 10000 year time scale that extends 4 km, though a few are concerned with materials ten times this age. The million years back to the start of the Ice Age in Britain, a 'recent' geological event, takes us 400km, more than the distance from London to Paris. To reach the age of dinosaurs, 100 million years ago, we would have to walk right around the Earth; to represent the whole span of geological time in these terms we would have to do the same thing 46 times.

The three photographs (figs 2, 3, 4) link these different time scales. The Palaeolithic handaxe from the Pleistocene of Suffolk, 200 000 years old, comes from the base of the archaeologists' scale, the top of the geologists'. The fact that it was discovered in the year of the French Revolution adds a historical dimension. The necklace was made by Bronze Age Man 4000 years ago from fossil sponges that are themselves 80 million years old. In the cave painting a Palaeolithic artist has captured the likeness of an extinct species, otherwise known only from fossil bones, teeth and tusks in Europe.

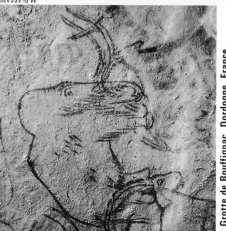

M PLASSARD

4 Grotte de Rouffignac, Dordogne, France

BM(NH)

3 Bronze Age necklace, Gravesend, Kent

BM(NH)

2 Palaeolithic handaxe, Hoxne, Suffolk

5 The scales of time and distance

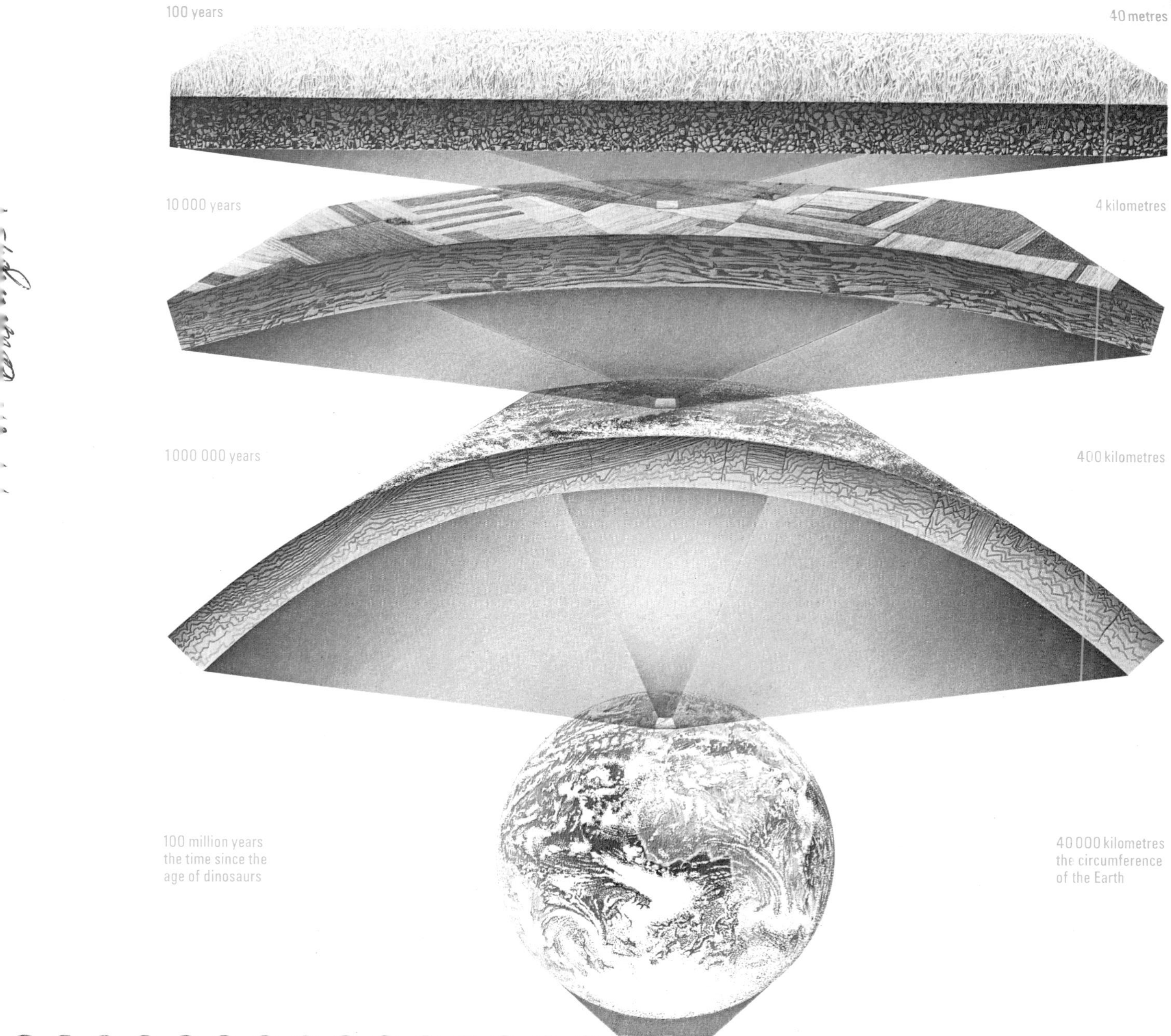

100 years

10 000 years

1 000 000 years

100 million years
the time since the
age of dinosaurs

the 4 600 million years since the formation of the Earth represented
by 46 globes, each containing the history of 100 million years.

40 000 kilometres
the circumference
of the Earth

400 kilometres

4 kilometres

40 metres

What can be learnt from rocks ?

Rocks exposed at the surface of the Earth or lying within reach of a drill are the source of most of our knowledge of Earth history. The geologist is able to extract information from even the most ordinary-looking rock sample.

Rocks are divided into three classes according to their origins: *igneous rocks*, such as basalt, crystallised from molten material either on the surface or below ground; *sedimentary rocks*, such as sandstone, formed from the weathering of existing rock, carried by water, wind or ice, and then laid down on land or under water; *metamorphic rocks*, such as marble and slate, formed by recrystallisation of any rock by heat and pressure, often deep underground. Ancient igneous and metamorphic rocks provide clues to the origin and early history of the Earth's crust; younger deformed rocks and igneous bodies allow an understanding of mountain chains and the overall structure and working of the globe; sedimentary rocks and the fossils they contain provide the evidence for our recon-

struction of the surface history of the Earth over the last 1000 million years. It is the layering or *stratification* of sedimentary rocks which gives its name to *stratigraphy*, the body of principles dealt with in the first half of this booklet.

Typical environments in which sediments are being deposited at the present time are shown in figure 6. Although rocks formed in shelf seas, in deltas, and by deep-sea turbidity currents make up the bulk of the geological record, all other environments make a contribution. Study of modern environments and the animals and plants associated with them allows the conditions of deposition of many ancient rocks to be identified. The mineral composition of a sedimentary rock, the shape of the grains that make it up and the gross features of its stratification will give clues to the type of rock that weathered to produce the sediment, the agent that carried the sediment from its source, and the climate and relief in the areas both of erosion and deposition. The total character of a sedimen-

tary rock, in mineral and fossil content and in texture, linked to a particular environment of deposition, is known as its *facies*. The sedimentary formations laid down in the different environments in figure 6 will, when turned into rock, each denote a different facies. Deposits of any shelf sea in the same situation as in the figure will be of the same general facies whatever their age. The concepts of facies and facies change are important when the age of a particular rock unit is being considered in detail (p 8).

Changes that take place between deposition and burial of a sediment and its appearance as a rock in a quarry or cliff must be understood if the original nature of the sediment is to be discovered. These changes, known as *diagenesis*, include compaction, recrystallisation, cementation, oxidation and reduction. A calcite mud buried on the floor of a warm shelf sea may appear in a road cutting as a flinty limestone, and a black mud may be altered into a pyritised shale.

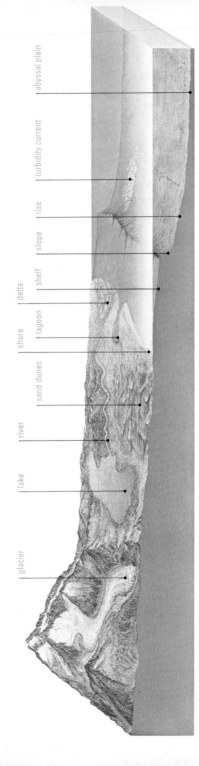

glacier · lake · river · sand dunes · shore · lagoon · delta · shelf · slope · rise · turbidity current · abyssal plain

continental environments   marine environments

6 Environments of erosion and sedimentation

9 Overturned strata, Keltie Water, Stirling

8 Unconformity, Kerrera, Argyllshire

7 Large-scale folds, Cariboo Mountains, Canada

Sediment, whether on a river bed or on the ocean floor, is generally laid down in flat sheets. Diagenesis emphasises the slight differences between successive sheets, so that sedimentary rocks often show very obvious layering. Individual layers are of the same age wherever they can be traced, and are younger than layers below, older than those above. These layers are often intricately folded and fractured by forces acting within the Earth's crust (fig 7) and they may be injected and baked by rising bodies of molten rock. The original layering may be masked as clays or shales are metamorphosed into cleaved slates and schists. These folded rocks may be lifted up to be exposed in an area of erosion, only to sink down and be covered by layers of flat-lying sediment, giving rise to a feature known as an *unconformity* (fig 8).

The geologist unravels these complications, attempting to discover the order of deposition of sediments and their original character, and giving a step by step account of the changes they have undergone. In areas of complex folding and faulting, layers may be turned upside down, so that the unravelling is only made possible by noting features of the sedimentary rocks which give a clue to their original way up. Ripple marks in sandstone or mudstone have sharp upward-pointing crests and rounded downward-pointing troughs (fig 10 shows the older surface of a vertical slab); where the size of particles in a sedimentary rock progressively changes, then the coarser layer is generally older than the finer (fig 11 is the right way up); and in rocks showing cross-bedding formed in dune or delta sands the younger beds always cut across the older (fig 12 is the right way up). Where a conglomerate in one rock unit contains distinctive pebbles from a nearby rock of unknown age, then the nearby rock is bound to be older than the conglomerate, whichever way up they are. Without the application of such criteria one could not tell that the horizontal beds shown in figure 9 have been overturned by folding so that the older now lie above the younger.

12 Cross bedding in dune sandstone, Arizona, USA

TAD NICHOLS

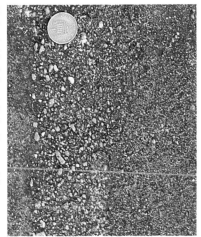

11 Graded bed, Stack Polly, Ross and Cromarty

IGS D2155

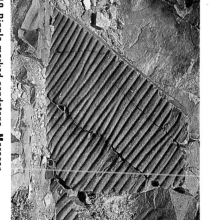
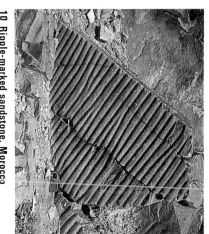

10 Ripple-marked sandstone, Morocco

F.W.DUNNING, IGS

Naming the layers

Aware of the amount of information that can be gained from studying rocks, the geologist sets out into the field to look at quarries, cliffs, road-cuttings and ditches, and to plot the position of the different layers he finds onto a map. The geological map he builds up will eventually show the pattern of rock outcrop and attitude of layers over his chosen area. If the rocks are well exposed, as in mountainous or rugged country, the map will be largely a record of observation, but more often the geological boundary lines have to be drawn out from only a limited amount of evidence. At the same time the geologist measures the thickness of strata exposed in cliffs and quarries to draw up a column section for the area. A satellite photograph of a tract of country (fig 13) will often indicate a rock pattern that can be seen in detail on the geological map (fig 14) and understood in relation to the column section (fig 15). From the sequence of beds in the column the geologist will reconstruct a story of changing environments of sedimentation and erosion. The map records such events as folding, faulting, and metamorphism which happened long after the sediment was laid down. These events can be discerned from a cross-section drawn through an area. To a geological eye, figure 16 reveals a story of deposition, folding, igneous intrusion, faulting, uplift, erosion, submergence, more deposition, uplift, and, at the present, erosion. The basic subdivisions of this local column must be defined and named.

They are known as *formations*, and are either single rock types or sets of distinctive rock types that can be mapped across country. A formation may be named after an area in which it is particularly well displayed: this area then becomes the *type locality* for the formation. Figure 17 shows part of the type locality of the Portland Limestone Formation of the British Jurassic. *Members* and *beds* are smaller units which are included within a formation, and *groups* and *supergroups* are the larger units. The boundary between two formations may be a sharp change in rock type and un-equivocal, or there may be a gradual change in character, in which case an arbitrary boundary must be drawn in the type locality. In a thick, monot-onous sequence, formations may be defined using features which are not obvious in the field, such as the presence of tiny mineral grains.

Once the local sequence has been established, the formations may be mapped out through the surrounding country. If there is no continuous outcrop, a separate exercise in naming must be carried out wherever they reappear, 50 or 100 km away. Many problems will arise, for when traced any distance the characteristics of a formation generally change. Beds of shale may appear within a sandstone, coal seams thin and disappear, and muddy limestones become sandy. To decide when a formation 'has changed so much that it needs a new name is always difficult. In general the larger units are recognisable over a wider area than the smaller.

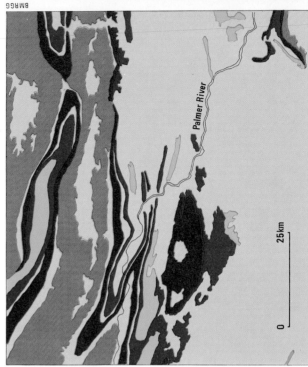

14 **Geological map of the Macdonnell Ranges**

BMRGG

Palmer River

0   25km

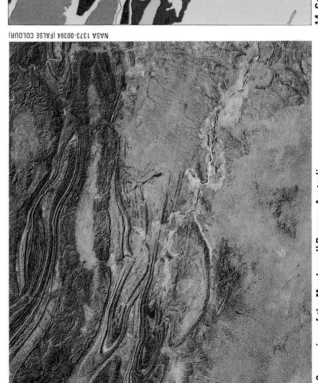

NASA 1373-00364 (FALSE COLOUR)

13 **Space view of the Macdonnell Ranges, Australia**

6

A particular type of rock-correlation is that in which the sequences in boreholes are compared. Although rocks can be examined and matched from chippings or cores, very useful information comes from the measurement of physical properties of the rock by instruments lowered down the borehole. Two of the most commonly used are radioactivity and electrical resistivity. One instrument can be lowered which measures the natural radioactivity of the rock layers, and another bombards the rocks with neutrons and measures the radiation that is produced. Resistivity is measured by running a pair of closely spaced electrodes down the borehole. Graphs produced in boreholes far apart can be compared and the different formations traced.

This whole procedure of naming, classifying and tracing stratified rocks is known as *lithostratigraphy*.

Many British rock units were named before the formal terminology described above came into use. The terms 'bed' and 'formation', as well as 'series' and 'division', are used indiscriminately, and many units have lithological designations such as 'clay', 'grit', or 'stone', which may not be truly descriptive. In addition, many geographic names, introduced long ago, refer to areas where the rock unit is now inaccessible. Thus, the Betton Beds and the Stapeley Volcanic Group are comparable units within the Llanvirn Series in Shropshire, the Millstone Grit of northern England contains more shale than grit, and now that the pits are closed there is not much Oxford Clay to be seen around Oxford.

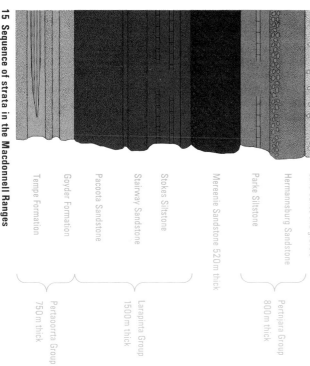

15 Sequence of strata in the Macdonnell Ranges

sand dunes and gravels

Hermannsburg Sandstone

Parke Siltstone — Pertnjara Group 800m thick

Mereenie Sandstone 520m thick

Stokes Siltstone

Stairway Sandstone — Larapinta Group 1500m thick

Pacoota Sandstone

Goyder Formation

Tempe Formation — Pertaoorrta Group 750m thick

16 A story of folding, intrusion and erosion

unconformity

17 Isle of Portland, Hampshire, type locality of the Portland Limestone

IGS A12236

Older or younger ?

In the same way that lithostratigraphy links layers of rock in different areas that belong to the same formation or bed, so *time-correlation*, or *chronostratigraphy*, links layers in different areas which were laid down at the same time. It is tempting to assume that a particular rock formation must be the same age wherever it is found, but this is often not so. In figure 6, if the shoreline were advancing out into the shelf sea, then the beach sand, a single unit in terms of lithostratigraphy, would be younger in the east than in the west, and would time-correlate with both dune sands and shelf sea deposits. A rock unit that is a different age in different places is said to be *diachronous*. Even where a formation does have a constant age it can rarely be traced more than a few hundred kilometres. Time-correlation must be worldwide if the geological history of the globe is to be worked out in detail. It can reveal whether a rock layer is older, younger, or the same age as another layer far away, but gives no clue to the actual age of the layers in years.

The most important method of time-correlation is using fossils; it is known as *bio-stratigraphy*.

It was the canal engineer William Smith (fig 18) who first used fossils to identify rock strata. He found that each of the rock units around Bath had characteristic fossils, distinct from those above and below (fig 19). He used this discovery to trace the units right across England without having to map every inch of the way. In 1815 he published the first large-scale geological map of any country in the world, *A delineation of the strata of England and Wales, with part of Scotland* (fig 20). This was followed in 1816 by the first part of *Strata identified by organized fossils*, a work in which typical fossils of each of the rock units he knew are illustrated on a suitably coloured background. His own collection of fossils, including most of the specimens depicted in the book, is preserved in the British Museum (Natural History) in London.

The Frenchman Alcide d'Orbigny, working 50 years after Smith, went much further. He recognised 27 successive fossil faunas in one part of the geological column which, as he believed each became entirely extinct as the next was created, provided an infallible means of time-correlation. He used fossils quite independently of rock units and, although his theoretical ideas were quite wrong, successfully correlated rock units over very great distances.

Acceptance of Darwin's theory of evolution in the 1860s discredited such correlation based on the concept of successive creations and led to the present less dogmatic approach. Two ideas form the basis of correlation using fossils today: first that all members of a species evolve together over their whole geographical range, so that evolutionary changes can be regarded as taking place at the same time wherever they occur, and second that evolution is a process which does not repeat itself, so that once a species or fauna has gone, it will never reappear. For a fossil to be useful in time-correlation it must be widely distributed in a variety of rock types, reasonably common and easy to recognise, and a member of a well defined, rapidly evolving lineage. No fossil satisfies all these requirements and all have their particular problems. The most useful are those like graptolites and ammonites which moved freely in the surface waters and are therefore found over wide areas in many different rock types. Less adequate are those like corals, gastropods, and bivalves, which evolved slowly and which were confined to a narrow range of environments. Widely used fossils, including some of the unfamiliar microscopic forms which are very important in borehole correlation, are shown in figure 21, overleaf.

18 William Smith (1769–1839)

19 Jurassic fossils collected by Smith

20 (opposite) Part of Smith's map of 1815, slightly reduced

The unit of time-correlation using fossils is the *zone*. This is a sequence of rock layers which contains a particular fossil or assemblage of fossils and is called after one of them, the *index fossil*. A zone may be defined by the range of a single fossil species, by the overlap in the ranges of two species, by the presence of an assemblage of different species, or by the local abundance of a particular fossil. These are the *range zone*, the *concurrent range zone*, the *assemblage zone*, and the *acme zone* respectively. Most zones are divided into subzones, each with its own index fossil. An example is the *Echioceras raricostatum* Assemblage Zone of the Lower Jurassic, which is recognised throughout Britain, France, Germany and Austria on the basis of ammonites. Four subzones of this zone are defined in Britain, although they cannot be recognised on the Continent.

Where the boundary between one zone and the next is marked by the evolution of one species into another then it will represent a true time plane. The boundary between two of the subzones of the *Peltura scarabaeoides* Assemblage Zone of the Upper Cambrian, marked by the evolution of one species of the trilobite *Ctenopyge* into another, is bound to be the same age whether it is found in Britain, Sweden, or eastern Canada. However, undisputed evolutionary boundaries such as this are much less common than those marked by the migration of species to or from the area. The base of the *Psiloceras planorbis* Zone of the Lower Jurassic is marked by the first appearance of this ammonite, which was migrating northeastwards from the Alps as the strata were being deposited. The boundary must therefore be older in Dijon, France, than in northwest Scotland, 1200 km further north. However, studies of the migration of living animals show that movements of hundreds of kilometres may be accomplished in a few centuries, so that, as even the shortest zone represents a period of many hundred thousand years, such variations in age may quite safely be ignored.

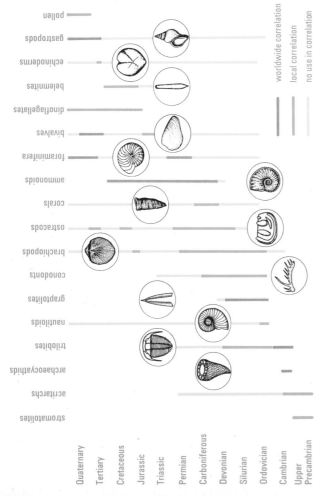

21 Fossils important in time-correlation

worldwide correlation
local correlation
no use in correlation

pollen
gastropods
echinoderms
belemnites
dinoflagellates
bivalves
foraminifera
ammonoids
corals
ostracods
brachiopods
conodonts
graptolites
nautiloids
trilobites
archaeocyathids
acritarchs
stromatolites

Quaternary
Tertiary
Cretaceous
Jurassic
Triassic
Permian
Carboniferous
Devonian
Silurian
Ordovician
Cambrian
Upper Precambrian

R.CASEY, IGS

22 Upper Myntlyn Beds, North Runcton, Norfolk

R.CASEY, IGS

23 Ryazan Beds, southeast of Moscow, USSR

10

Few of the fossils used in correlation are equally abundant in all types of sedimentary rock. Just as most modern animals only live in a certain range of environments, so most fossils are only found in rocks of certain facies. Rocks of the Devonian System comprise three main facies in Europe, each with different fossil faunas and different zonal schemes. The Old Red Sandstone, laid down in lakes and estuaries, is zoned by fossil fish; the Rhenish facies, using brachiopods and corals; the Hercynian facies, laid down in deep, muddy water, is zoned on the basis of ammonoids. Discovery of Old Red Sandstone interbedded with Rhenish facies rocks in north Devon, and rare Hercynian ammonoids in sandstones has allowed some correlation between the three schemes.

Even within rocks of a single facies, few index fossils are found all over the world. Just as modern animals and plants are grouped into a number of great *faunal provinces* based on the major continents and oceans, so most fossils are found only within a certain area. Two provinces existed in the seas of the northern hemisphere from 150 to about 100 million years ago. In the north was the Boreal Ocean, at first largely enclosed by land, where ammonites and other marine animals evolved rather differently from those in the Tethyan Ocean to the south. Although temperature may have played a part, it was the geographical isolation that was crucial. Ammonites typical of a Tethyan Province are found in southern USA, around the Mediterranean, and in Tibet. Boreal Province ammonites of the same age are found in north-ern Europe and Russia. In between these two great provinces are found the ammonites of southern Britain and France. Different zonal schemes are used in these three areas, and once again, correlation depends on finding occasional specimens 'out of place'. Figures 22 and 23 give an example of time-correlation of lowermost Cretaceous rocks within the Boreal Province. The ammonite *Surites* is found at both localities, 2500 km apart.

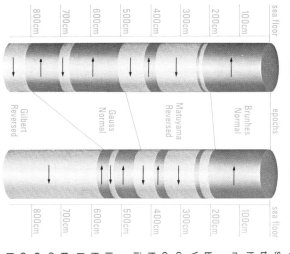

25 Magnetic correlation of oceanic cores

24 Portaskaig Boulder Bed, Islay

F.W.DUNNING, IGS

Using rocks. Fossils are the most important means of time-correlation, but not the only one. Other methods not only confirm the evidence of fossils, but also enable unfossiliferous strata to be time-correlated to some extent. Any thin bed of rock that can be shown to be the same age throughout its extent may be used in time-correlation. Such a 'key' bed is said to be *isochronous*.

Beds of lava or volcanic ash are isochronous, and can provide useful correlation where they are interbedded with sedimentary rocks: ash beds have been traced 50 km in Wyoming and Montana, USA, and have proved the diachron-ism of a Cretaceous shale unit. Beds of evapor-ite, minerals laid down by evaporating sea-water, although varying slightly in age from the edge to centre of the basin of evaporation, are nevertheless useful time-markers in unfossi-liferous sediment. Glacial deposits have been used for long-range correlation in the upper Precambrian of northwest Europe, as sediment laid down during the few million years of an ice age is effectively isochronous when compared with the span of Precambrian time; the Porta-skaig Boulder Bed, a fine-grained deposit con-taining boulders dropped by melting icebergs, has equivalents from the west coast of Ireland to northern Norway (fig 24).

Many of the Pleistocene sediments laid down by melting ice show fine annual banding. Variations in the thickness of these bands allows correlation between different exposures, and detailed chronologies of the last 10000 years have been built up in Scandinavia and America in this way.

Sediments from the ocean floor can be corre-lated from their palaeomagnetism. Every few hundred thousand years the direction of the Earth's magnetism changes, the north magnetic pole becoming the south and *vice versa*. Evi-dence of these changes is preserved in the oceanic sediments, allowing an accurate time-correlation (fig 25). The sedimentary record on land is too fragmentary for this method to be applied widely.

The stratigraphic scale

When rock units have been named and classified into groups and formations, and correlated using fossils and key beds, it remains to define terms which will indicate the age of a rock layer irrespective of its fossil content. One set of terms is used to define periods of time, and a related set to refer to the rocks laid down within them. The smallest time period is the *chron* during which a *chronozone* of rock is laid down; the larger pairs of terms are *age/stage*, *epoch/series*, and *period/system*. Thus the Triassic System of rock was laid down during the Triassic Period of time, and the Carnian Stage during the Carnian Age.

The chronozone is defined and named in a particular exposure or borehole, which then becomes its type section. The stage may have its own type section or may be defined in terms of the chronozones that make it up: stages have geographical names mostly ending in '-ian'. Systems have no type sections and their names, often of some antiquity, relate to the area where they were first described, to a characteristic rock type, or to some feature of their fossil fauna.

In the present state of stratigraphic knowledge only the twelve systems can be recognised all over the world. Forty years ago there was little agreement over naming and boundaries even at this level, but today only few uncertainties remain. The most important of these are the placing of the boundary between the Cambrian and Ordovician systems, the use of the single Carboniferous System in Europe as against the Mississippian and Pennsylvanian systems in North America, and the choice of a name – 'Quaternary', 'Anthropogene' or 'Pleistogene' – for the most recent system.

Stages can be recognised over very large areas, but the problems of facies and faunal provinces (p11) hinder their worldwide recognition; marine and freshwater facies of the Devonian have different stage names, as do the Boreal and Tethyan provinces of the Cretaceous. As correlation improves over the years these local schemes will hopefully be replaced by a single standard. The lower Silurian of Britain for example was named 'Valentian' in its deep-water, graptolitic facies, and 'Llandoverian' in its shallow-water, shelly facies. These two facies can now be correlated in sufficient detail for the latter name to be applied to both. A stage that has been traced throughout the world is the Callovian of the Jurassic System; figures 26, 27, 28, show it in three widely separated localities.

Little attempt has yet been made to use the chronozone for long-range correlation; indeed it is rarely to be found in the geological literature. In its type section it usually coincides with a fossil zone, and therefore can automatically be recognised throughout the area of that zone. All the skills of the stratigrapher will be needed to carry the chronozone into unfossiliferous sequences or those having different schemes of zones.

The different types of stratigraphic subdivision described above may now be summarised. 'Lithostratigraphic' units (groups, formations, mem-

26 Callovian mudstones, Jameson Land, Greenland

27 Callovian limestones, Verona, Italy

28 Callovian shales, Andes Mountains, Argentina

bers and beds) are based on rock type; they can be mapped in the field, are often diachronous and of small geographical extent. 'Biostratigraphic' units (zones) are based on fossils; they are isochronous, are of wider extent than most lithostratigraphic units, but are seldom worldwide. 'Time-rock' units (system, series, stage, chronozone) are based on time-correlation from a type section; they are isochronous and, potentially, worldwide (fig 29).

Enormous, perhaps insuperable, difficulties lie in the way of an all-embracing, universally acceptable stratigraphic classification of even the part of the geological column since the start of the Cambrian. These partly stem from features inherent in the rock record, and partly from a reluctance to give up well-tried standards. The scheme outlined above is only one of a number put forward over the last hundred years, each of which has had its critics and supporters.

Supporters of these schemes have stressed that they are the only possible basis for communication between workers in different parts of the world, and that without them there can be no real accumulation of geological knowledge. They insist that the general benefit of these schemes outweighs the temporary inconvenience that goes with any change of classification.

Critics of the trend towards formalisation point out that by forcing the stratigraphic column into a straitjacket of theoretical categories order may be gained, but at the expense of a close approach to reality. They stress that the existing nomenclature, a complex of terms often based on rock type, fossil content, and time, provides a meaningful and natural classification of the geological column which should not be swept away. They admit that a formal scheme would be valuable when approaching an unknown area, or perhaps even when dealing with strata that are only moderately fossiliferous. But in highly fossiliferous divisions, such as the Jurassic of Europe, there seems to them little point in erecting a new classification based on rock type alone which takes no account of the abundant and distinctive fossils.

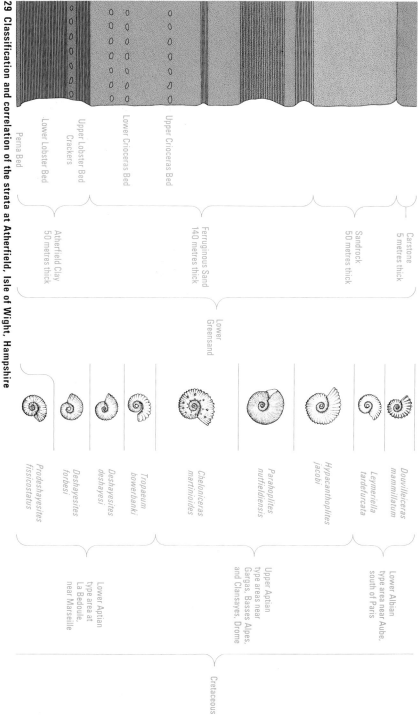

29 Classification and correlation of the strata at Atherfield, Isle of Wight, Hampshire

Dating the layers

30 Lord Kelvin (1824–1907)

31 Arthur Holmes (1890–1965)

32 Mass spectrometer for isotopic dating

A 17th century scholar, Archbishop Ussher of Armagh, worked out from biblical evidence that the world was created within six days in 4004 BC. This date, printed as a marginal note in the Authorised Version of the Bible, was accepted by many scientists into the 18th century and by the general public until the beginning of the 19th. Even today this view has a few adherents.

The middle of the nineteenth century saw the start of a controversy between geologists, led by Charles Lyell who believed that the Earth was so old as to be infinitely old for all practical purposes, and the physicists, led by Lord Kelvin (fig 30) who calculated that the time needed for the Earth to cool to its present state from its molten origin was about one hundred million years. Many geologists began to think critically about the age of the Earth and, basing their calculations on present-day rates of erosion and the build-up of salt in the seas, came up with figures not too far from those of Kelvin. In the 1880s, however, he reduced his figure to between twenty and forty million years. The geologists felt certain that the complex story of geographical and biological change that they had worked out from the rocks could not be compressed into such a short time span, but they could find no flaw in the intricate calculations of their opponents.

Radioactivity, the emission of particles and radiation by the atoms of certain elements, was discovered by Henri Becquerel in 1896. When in 1903 it was found that radioactive materials produce heat energy, it was realised that the Earth is not a simple cooling body, but has an internal heat source which slows its overall cooling. The physicists, having admitted that their estimates of the age of the Earth must therefore be much too small, went on to use radioactive elements as geological clocks. They found that as these elements emit radiation and particles a more stable element is ultimately formed, and, because the rate at which the element is changing at the present day can be measured, the proportion of radioactive 'parent' to the stable 'daughter' element gives the age of the mineral containing them. The first crude dates, published in 1906, gave 2000 million years as the age of a radium mineral and 400 million as the age of a lead ore. It was Arthur Holmes (fig 31), then a student at Imperial College, London, who became the chief proponent of this method of dating and who, in 1913, published a classic book on the subject, *The Age of the Earth*. Many geologists remained suspicious of the new methods, still producing estimates based on salinity and the rates of geological processes. However, by the early 1930s the critics were silenced and the method was generally accepted. The major divisions of the Phanerozoic were dated, with 500 million years accepted as the date of the start of the Cambrian Period. The age of the Earth was tentatively suggested as 2000 million years.

What is radioactivity?

Atoms of a radioactive element have unstable nuclei that progressively 'decay' to a more stable form with the emission of radiation and particles. A composite particle is emitted in α-decay, an electron in β-decay, and radiation in the decay known as 'electron capture'. Stabilisation is not always a simple one-step process. Uranium-238, for example, goes through many changes before the stable form, lead-206, is reached. The number refers to the number of particles in the nucleus; forms of the same element having nuclei with different numbers of particles are known as *isotopes* of the element, hence the expression *isotopic dating*.

Some isotopes decay rapidly and are highly radioactive; others decay slowly and are less radioactive. There is no way of predicting at what moment any one nucleus will decay, but in a mineral specimen containing billions of atoms the time needed for any percentage of the unstable nuclei to decay is a very exact quantity. The 'half life' of an element is the time taken for half of its nuclei to decay. A highly radioactive element may have a half life of a fraction of a second, whereas others are millions of years long. Thorium-231 has a half life of 25.6 hours, so half of a pure sample will decay in this time, half of the remainder in the next 25.6 hours, the radioactivity getting weaker all the time but never quite disappearing.

Before any age determination can be made, the half life of the isotope concerned must be accurately measured in a nuclear physics laboratory. Pure samples are prepared and the very low levels of radioactivity are picked up on highly sensitive equipment. The more accurately the half life is determined the more accurate will be all the age determinations based upon it. Uncertainties in the half life of rubidium-87, figures ranging from 47 000 to 52 500 million years having been used in recent years, have introduced errors of up to 10 per cent in published dates. Another preliminary step is to establish that the change in the proportion of parent to daughter elements in the sample is entirely due to radioactive decay. If the daughter is a gas it may diffuse away and the sample will give an age which is too small, or if there was daughter element present in the sample before decay began it must be identified and considered in the final calculation, otherwise the result will be too great.

Uncertainties such as errors in half life determination and in the analysis are expressed in a plus or minus figure, for example 500 ± 15 million years.

Once the half life and the proportion of daughter element to parent have been measured, the date to be given to the sample may be found by calculation.

Parent isotope and type of decay	Percentage of parent in normal element	Half life	Daughter isotope	Geological application
uranium-238 α and β decay	99.3	4510 million years	lead-206	uraninite and pitchblende in uranium ores, zircon and monazite in granitic rocks more than 50 million years old
uranium-235 α and β decay	0.7	713 million years	lead-207	uraninite and pitchblende in uranium ores, zircon and monazite in granitic rocks more than 20 million years old
thorium-232 α and β decay	100	13 900 million years	lead-208	
potassium-40 electron capture	0.019	1300 million years	argon-40	micas, potassium feldspars and pyroxene in volcanic and granitic rocks, glauconite in glauconitic sandstone, and whole volcanic rocks more than 100 000 years old
rubidium-87 β decay	27.2	47 000 million years	strontium-87	micas and potassium feldspars in granitic rocks, glauconite in glauconitic sandstones, and whole igneous, sedimentary, and metamorphic rocks more than 10 million years old
carbon-14 β decay	less than 0.000 0001	5570 years	nitrogen-14	fossil wood, shell and bone, fabric, pottery and ash from archaeological sites between 70 000 and 1000 years old

33 Radioactive decay series and their geological applications

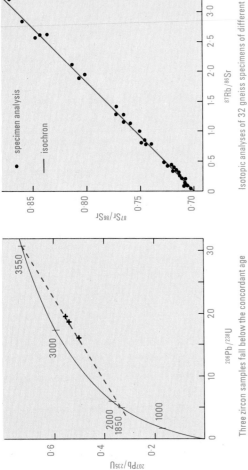

specimen analysis
— isochron

Isotopic analyses of 32 gneiss specimens of different compositions from the Godthaab area in West Greenland plotted on a graph. The slope of the line, known as the isochron, gives the age of the gneiss, 3750±80 million years.

Three zircon samples fall below the concordant age pattern expected for the rock body and suggest dates between 2640 and 3280 million years. The intersections of their straight line on the curve give the true age, 3550 million years, and the date of lead-loss, 1850 million.

34 Discordant age pattern

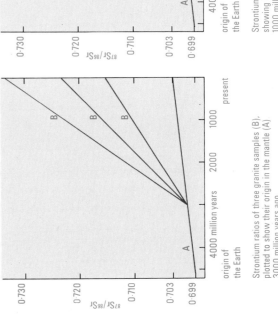

35 Age-dating by isochron

Strontium ratios of another granite (C) showing how it formed from a type B granite 1000 million years ago.

Strontium ratios of three granite samples (B), plotted to show their origin in the mantle 3000 million years ago.

36 Ratios of strontium isotopes reveal the date and place of origin of granite bodies

Techniques. Chemical methods are used to extract the elements required from the rock or mineral sample. The mixture of elements is analysed in a mass spectrometer (fig 32), which can measure small traces of isotope with great accuracy. The technique of 'isotope dilution', addition of the mixture being analysed to one of accurately known composition, makes the difficult procedure a little easier.

Because each dating method has its own problems, two or more are generally used on each sample. Minerals that contain potassium, such as mica, dated by the potassium/argon method, also contain traces of rubidium, and can be checked by the rubidium/strontium method. If different methods give the same date then the specimen is said to show a *concordant* pattern. If the dates are different (*discordant*), then important information on the geological history of the sample may be deduced (fig 34).

An important way of minimising analytical error, particularly in the rubidium/strontium (Rb/Sr) method, is by use of the *isochron*. Rocks of the Earth's crust contain two isotopes of strontium, 87 and 86. Because of the very slow production of strontium-87 by the decay of rubidium-87, the ratio $^{87}Sr : ^{86}Sr$ has increased slightly since the formation of the Earth. The average in the mantle today is 0.703; 4600 million years ago it was 0.699 (fig 36, line A). If different parts of a rock were analysed at the time of its formation, their rubidium content would vary but their $^{87}Sr : ^{86}Sr$ ratio would be constant. If these analyses were plotted on a graph of $^{87}Sr/^{86}Sr$ against $^{87}Rb/^{86}Sr$ a horizontal line would be the result. As time passes the samples would decrease in rubidium and increase in strontium through radioactive decay. This would happen most rapidly in the samples richest in rubidium, so that the analyses would begin to plot onto a sloping straight line, the *isochron*. The greater the age, the steeper would be the slope of the line (fig 35).

Figure 36 shows how a study of the change in the $^{87}Sr : ^{86}Sr$ ratio can give clues, not only to the age, but to the place of origin of a rock body.

Applications. The stratigraphic methods outlined in the first part of this booklet are of little use in the classification and correlation of the Precambrian. The few fossils that have been found are of little or no stratigraphic value, and there was no way of relating Precambrian rocks in different parts of the world or of setting up a widely applicable scheme of classification. Lithological mapping did little more than bring local order to the great metamorphic terrains and piles of unfossiliferous sediment. Isotopic dating reveals not only the great length of Precambrian time, but that rocks in the great shield areas of Africa, Asia and Canada have been affected by a number of mountain-building episodes, each characterised by widespread metamorphism and igneous intrusion. In the Canadian shield for example many rocks give dates around 2500, 1800 and 1000 million years. Broadly similar dates may be recognised in many other parts of the world, and are taken as the boundaries of major time divisions within the Precambrian.

Dating the divisions of the Phanerozoic (post-Precambrian) stratigraphic column is another important application of isotopic dating. The only sediments that can be dated directly are those in which a radioactive mineral is formed during diagenesis of the sediment, such as the rather uncommon illite shales and glauconitic sandstones; other sediments give only the age of the parent rock from which the mineral grains that make them up are derived. Where lavas or volcanic ashes are interbedded with a sediment of known stratigraphic age (fig 38), then a date may be given to that stratigraphic division. Where an igneous rock intrudes one sedimentary unit and is blanketed by another, then the sediments may be dated from the igneous rock by inference. The rarity of such cases, together with the analytical error inherent in age determination, mean that isotopic ages are unlikely to rival or replace fossils as the most important means of Phanerozoic correlation. Considerable progress has been made in dating down to the level of stages.

Use of radioactive carbon-14 has allowed the dating of events of the last 50000 years, a time interval too short for correlation using fossils, although full of interest for both geologist and archaeologist. Carbon-14 is being produced all the time from nitrogen-14 by cosmic bombardment in the upper atmosphere, and mixes so that a tiny proportion of all the carbon in the air is radioactive. When carbon is taken up from the air by an animal or plant and incorporated into bone or woody tissue, the proportion steadily decreases as the radiocarbon decays to stable nitrogen-14. The age can be calculated if the very weak emission of β-particles is measured using geiger tubes or a scintillation counter. Natural radiation from the atmosphere must be carefully excluded before measurements can be made. Wood, peat, shell, bone, hide and fabric may all be dated in this way (fig 39). Corrections are necessary to all carbon dates as it is now known that the amount of carbon-14 in the air has not remained constant but has changed through the centuries.

37 Precambrian terrain, Scourie, Sutherland

J.M.PULSFORD, IGS

38 Lava below Devonian sandstones, Hoy, Orkney

W.MYKURA, IGS

39 Dated shells from South Shian, Argyllshire

IGS

17

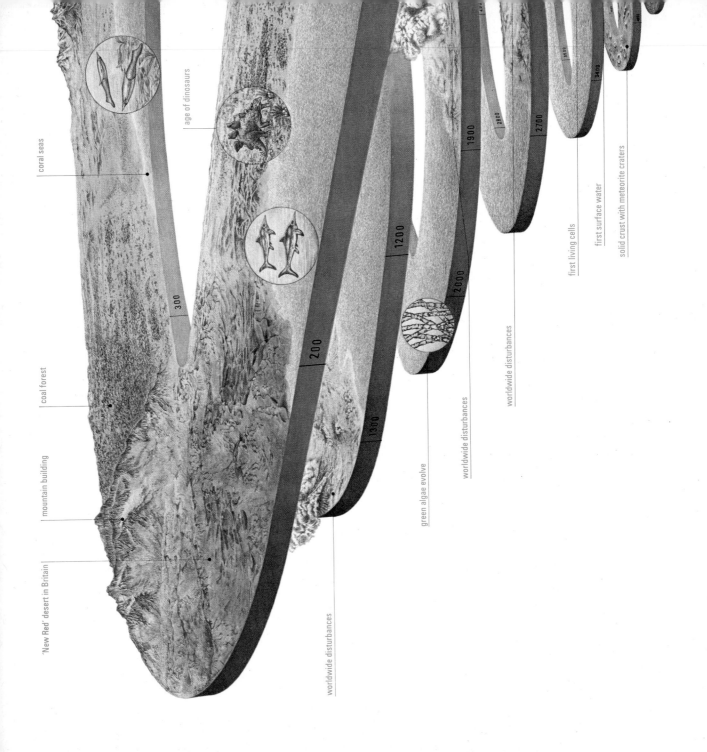

coral seas

age of dinosaurs

'New Red' desert in Britain

coal forest

mountain building

300

200

1200

1300

2000

1900

2800

2700

1900

green algae evolve

worldwide disturbances

worldwide disturbances

worldwide disturbances

worldwide disturbances

first living cells

first surface water

solid crust with meteorite craters

3400

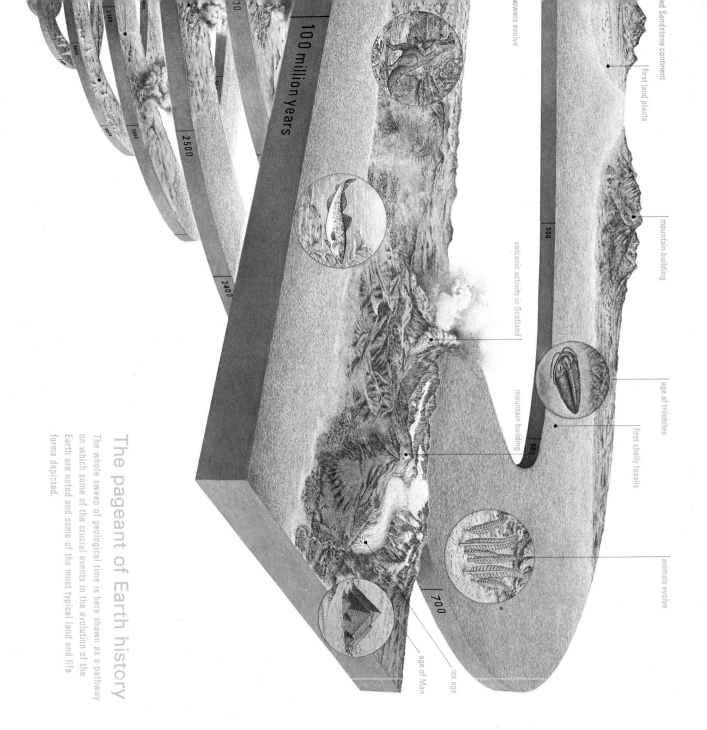

The pageant of Earth history

The whole sweep of geological time is here shown as a pathway on which some of the crucial events in the evolution of the Earth are noted and some of the most typical land and life forms depicted.

100 million Years

1800

2500

2400

2200

2100

Sandstone continent

first land plants

mountain building

500

volcanic activity in Scotland

600

mountain building

age of trilobites

first shelly fossils

animals evolve

700

ice age

age of Man

owers evolve

How old is the Earth?

Rocks up to 2800 million years old are common in the great Precambrian shield areas, but rocks dating back 3500 million years or more have been found only in a few places. The Morton Gneiss from Minnesota, USA, the ultrabasic rocks from the Barberton Mountain Land of Swaziland, the Warrawoona Group from Western Australia, and gneisses from west Greenland and the Antarctic are all examples of these very ancient rocks. The oldest rock in the world is currently a pebble of volcanic ash from a conglomerate near Isua, West Greenland, which has yielded U/Pb dates of 3824 ± 12 (fig 40a). Other rocks close to this age in the same area are the Amitsoq Gneiss (fig 40b) and a banded iron formation (fig 40c). An ancient Antarctic rock is also shown in the photograph (fig 40d). These rocks prove that the Earth had a solid crust at least 3850 million years ago.

Meteorites (fig 41) have been dated by U/Pb and Rb/Sr methods to about 4600 million years. They are thought to originate in the belt between the orbits of Mars and Jupiter and to be the fragments of asteroidal bodies which disintegrated long ago. Since the uniform orbits of the planets suggest that the Solar System had a single origin, it is likely that the Earth, as well as these asteroidal bodies, formed 4600 million years ago. The oldest rocks so far recovered from the Moon are coarsely crystalline, quite different from the less ancient basalt lavas and breccias (fig 42). They seem to be part of the original crust of the Moon and have been dated at 4600 million years. As there is every reason to believe that the Earth and Moon formed together, this is further indirect evidence for the age of the Earth. Direct evidence comes from the change in lead isotope ratios through time. Lead isotopes 206 and 207 are steadily produced by the decay of uranium, and the ratio of their abundance to lead-204 is increasing. Formation of lead ore bodies at different times in the past removed lead from the uranium and 'froze' its isotopic composition. Their lead isotope radios can be plotted against time on a graph (fig 43); if this graph is extrapolated back in time, the average isotopic composition of lead in iron meteorites is reached at 4600 million years, suggesting once again that the Earth and the asteroidal bodies had a common origin at that date.

This evidence and much more indicates that by about 4600 million years ago a vast disc-like cloud of gas and dust had condensed to form the star/planet system we know as the Solar System, and the Earth itself was formed.

Looking further back in time, some estimate can be made of the age of our galaxy with its hundred thousand million stars. Most stars seem to have a life cycle of steady size and brightness followed by rapidly in-creasing size and decreasing brightness and a final 'white dwarf' stage. Calculations involving the luminosity and mass of the oldest stars visible from the Earth show them to be about 11 thousand million years old, which is probably the minimum age of the galaxy.

IGS T483

IGS T384

40 The oldest rocks on Earth

IGS T389

41 Stony meteorite, Barwell, England

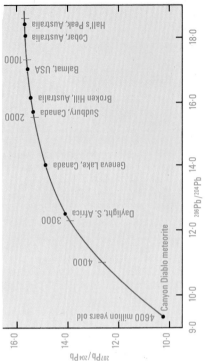

42 Moonrock: basalt on anorthosite

43 Lead ratios and the age of the Earth

Primeval Earth

Condensation of part of the vast cloud of cold dust and gas that gave rise to the Solar System initially formed a molten Earth surrounded by a thick and dense atmosphere of cosmic gases. At a very early stage this atmosphere, made up largely of carbon dioxide and carbon monoxide, was stripped away, leaving the planet unprotected from meteorites, some of them enormous, which must have plunged into the molten mass. As the globe slowly cooled, crystallisation of minerals began. Iron globules sank to the centre where the core of the Earth was forming, and 'basic' rocks such as gabbro and anorthosite – feldspar rock – began to make a crust over the surface. As the molten rock solidified, gases, including water vapour, carbon dioxide and nitrogen, were given off and began to build a new atmosphere around the Earth. Water vapour condensed and fell as rain, starting the processes of erosion and sedimentation. This was the stage when the Amitsoq Gneiss was being metamorphosed, 3750 million years ago.

Small rafts of low-density granitic crust grew as the basic rocks were repeatedly cracked and remelted. Moved over the globe by convection currents in the plastic mantle, these continental rafts would occasionally collide and join together. Large bodies of water, the first oceans, collected in low-lying areas between the continents. Primitive living cells had developed in warm volcanic pools at a very early stage, and 2000 million years ago plant cells capable of photosynthesis evolved and started to add oxygen to the atmosphere. The major components of the Earth were all established by this time.

The three illustrations (figs 44, 45, 46) from a painting in the Geological Museum, London, show an artist's impressions of the molten surface of the globe as the first solid rock began to form, the solid crust with meteorite craters and a molten pool where one meteorite has punched through the Earth's thin skin, and a view of an early ocean.

44 Semi-molten surface of the Earth 4500 million years ago

B.EVANS/IGS T191

45 Solid crust with meteorite craters 4000 million years ago

B.EVANS/IGS T385

46 An early ocean 3800 million years ago

B.EVANS/IGS T484

Age patterns of the Earth

When mapped on a grand scale, the rocks of both continents and ocean floors show age patterns which give clues to their origin.

Continents are seen to consist of ancient centres, or 'cores', surrounded by successively younger tracts, something like the rings of a tree. The outermost tract may be a mountain range still in the course of formation. North America shows the pattern particularly clearly (fig 47), though it can also be traced in South America, Africa and Asia. It is not known for certain whether the younger tracts rose from the mantle at the dates indicated, or whether they are more or less the same age as the central cores, having been reworked in continental collisions. In northwest Scotland the junction of two such tracts can be studied, and structures such as basalt dykes traced from the older into the younger, where they are progressively deformed and ultimately obliterated. This suggests reworking. Application of the strontium ratio method (p16) on the Nûk gneisses of western Greenland on the other hand, shows that they are not a reworked extension of the nearby Amitsoq Gneiss, 3750 million years old, but rose from the mantle at a later date. It seems likely that continents have grown throughout geological time, but most quickly at a very early date. Over half of the present volume of continental crust seems to have been in existence 2500 million years ago. Since then most continental growth has probably taken place along subducting plate margins.

Age patterns on the ocean floor are much simpler than those of the continents. The age of the basalts which underlie the layers of soft sediment that make up much of the ocean floor increases on either side of the great underwater mountain ranges known as oceanic ridges. Even at the edge of an ocean basin, far away from the ridge, few basalts are more than 200 million years old. Great faults cut through the oceanic crust offsetting the line of the central ridge. These *transform* faults are active as the sea floor spreads and are sites of regular earthquakes. Only the largest are shown in fig 48. These facts support the concept of sea-floor spreading, the idea that the ocean ridges are sites of slowly upwelling currents in the Earth's mantle that spread the sea floor apart as new crust is injected (fig 53). The steadily cooling crust moves away from each side of the ridge, forming a part of the ocean floor for many millions of years until eventually it sinks back into the mantle in areas known as subduction zones, the sites of deep ocean trenches and island arcs. The continents, being granitic and having a lower density than the basaltic oceanic crust, are carried with the moving sea floor but are not subducted. The Atlantic Ocean (fig 48) has opened over the last 200 million years and is still widening at the rate of a few centimetres each year. The theories of sea-floor spreading and plate tectonics are fully explained in a companion booklet, *The story of the Earth*.

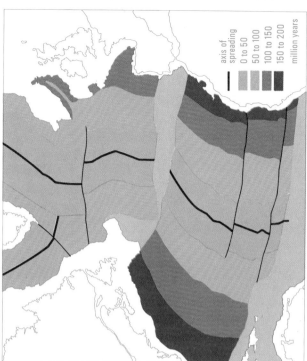

48 Pattern of ages in the north Atlantic

axis of
spreading
0 to 50
50 to 100
100 to 150
150 to 200
million years

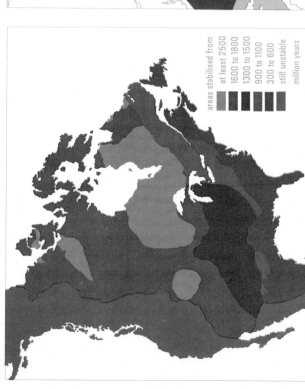

47 Pattern of ages in North America

areas stabilised from
at least 2500
1600 to 1800
1300 to 1500
900 to 1100
300 to 600
still unstable
million years

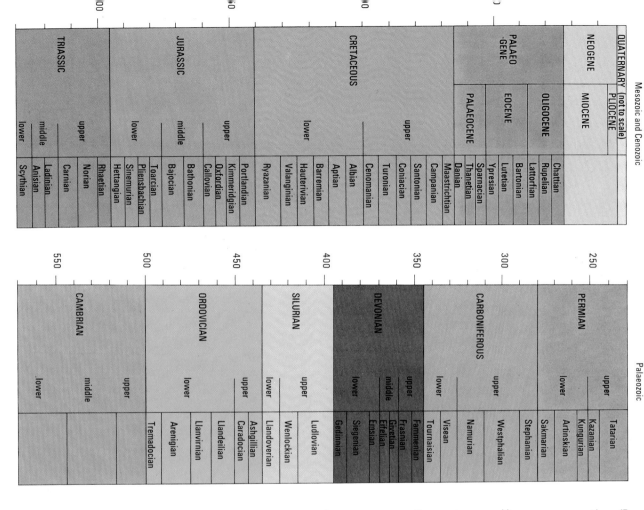

Mesozoic and Cenozoic

Palaeozoic

Precambrian

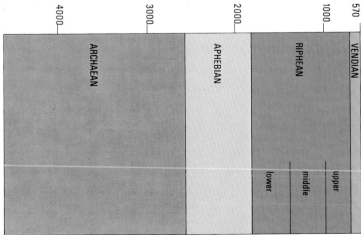

Geological column

This table shows the subdivision of the geological
column into systems, series, and stages. Dates, in
millions of years, are taken from the detailed geological
time table by Van Eysinga (1975). Names given are
those commonly in use for marine rocks in northwest
Europe. The fifteen stages of the Quaternary and the
eight of the Neogene are omitted for lack of space, as
are the modern stage names now accepted for parts of
the Ordovician, Silurian, and Carboniferous. In these
Palaeozoic systems the more familiar 'series' are listed.
The subdivisions of the Precambrian are based entirely
on isotopic dates and include both stratigraphic rock
units and metamorphic complexes. Neither the names
nor the dates on this table will be acceptable to every
geologist, as there are areas of disagreement affecting
nearly every part of it. The steady accumulation of new
data slowly reduces the size of these areas.

The ever-changing world

Geological processes in the past

In the 19th century there were two schools of thought about the way geological processes acted in the past.

One school, led by Charles Lyell (fig 49), believed that the geological processes at work on the Earth today have acted with the same intensity from the earliest times to the present. Knowing how slowly and gently the processes or erosion and sedimentation, uplift and down-warp work at the present, they believed that, to allow for the vast geographical changes that have taken place, the Earth must be immensely old. They insisted that the geological record gave no evidence of the formation of the Earth, but only of uniformity as far back as they could trace it. Their prize piece of evidence was the

Temple of Jupiter Serapis near Naples (fig 50) where three fragile columns remain standing in spite of considerable changes in sea level over the last 2000 years. This was seen by the 'uniformitarian' geologists as proof of the slow and gentle nature of geological processes.

The other school included geologists who thought that the geographical changes were too great to be accounted for by known geo-logical processes. They believed that the pro-cesses now operating must have acted with greater energy in the past, and that there must have been others now unknown. This speeding up of geographical change meant that, for them, the Earth was old but not immensely old. They also believed that there was evidence of the

Earth's origin as a molten globe, and that the decrease in energy of its processes is related to its cooling. One of their best pieces of evidence was the presence of great erratic boulders on the plains of northern Europe, far from any out-crop of similar rock (fig 51). They believed that only a catastrophic flood, greater than anything recorded in historic times, could explain their distribution. This was the 'catastrophic' school of geology.

Modern geological thought contains ele-ments of both these philosophies. The present is still used as a key to the past in the recon-struction of ancient environments, but it is realised that the story of the primeval Earth cannot be told in strictly uniformitarian terms.

51 Erratic boulders, Llanberis, Wales

50 Temple of Jupiter Serapis, Naples, Italy

49 Charles Lyell (1797–1875)

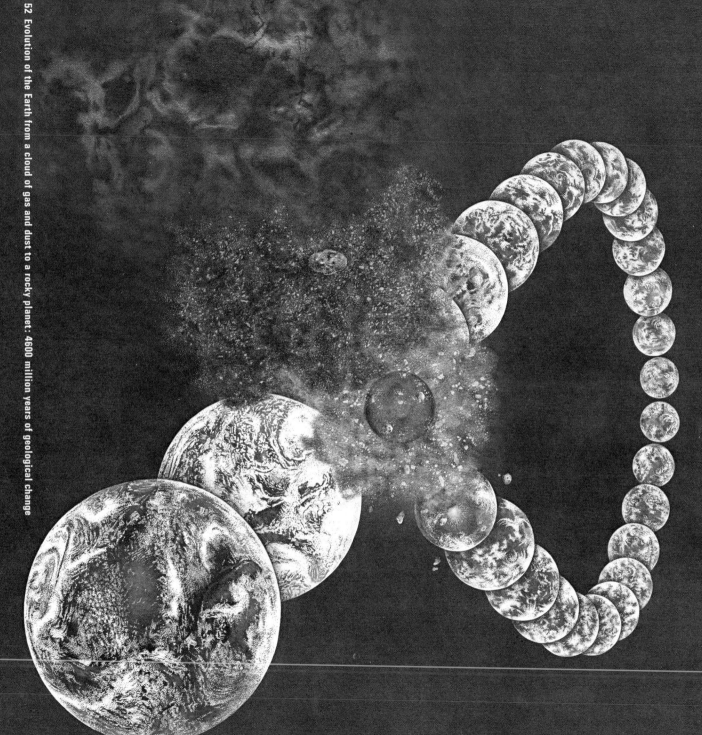

Geography and climate in the past

Geography, more particularly the distribution of land and sea, has been changing throughout Earth history under the influence of three geological processes. The slowest of these is the steady growth of continental crust by separation of granite from the mantle, which was accompanied by the condensation of water vapour from volcanic gases. The second process is the spreading of the sea floor from the lines of the ocean ridges and the drifting of the raft-like continents at rates of one or two centimetres each year (fig 53). Evidence of continental drift comes from the fit of landmasses that were once united, such as Africa and South America (fig 54), and from the apparent movement of the north pole when the magnetism of rocks of different ages is plotted (fig 55), as well as from the structure and age patterns of the ocean floor itself (fig 48). The positions of the continents at four times in the Phanerozoic are shown overleaf. The third process is a changing sea level, either due to variations in the amount of water in the ocean basins, or to a vertical movement of the crust. Mountain building during the Phanerozoic has steadily reduced the area of shelf sea by thickening the continental crust. Formation of a new stretch of oceanic ridge would raise the sea level and increase the area of shelf sea. Build-up of polar ice cap during an ice age lowers the sea level. Even the deposition of thick layers of sediment, as in the Mississippi delta, causes a local geographical change by depressing the continental crust.

Our ability to draw palaeogeographic maps depends on the distribution at the surface and in boreholes of rocks of the date required. Usually our knowledge is incomplete and patchy, but occasionally enough evidence is preserved in the rocks of a limited area to build up a detailed picture of the ancient geography. Four examples from Britain and France are shown overleaf (figs 61–64).

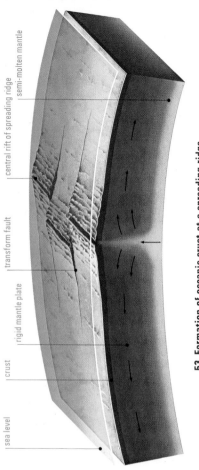

central rift of spreading ridge
semi-molten mantle
transform fault
rigid mantle plate
crust
sea level

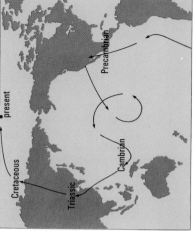

53 Formation of oceanic crust at a spreading ridge

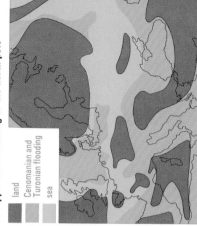

Cretaceous
Triassic
Cambrian
Precambrian
present

55 Apparent wandering of the north pole

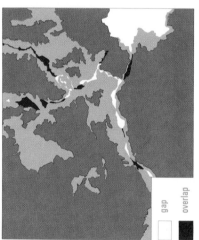

gap
overlap

54 Fit of the northern continents

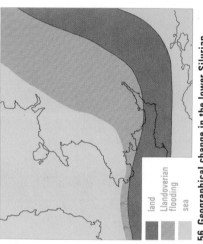

land
Cenomanian and Turonian flooding
sea

57 Geographical change in the upper Cretaceous

land
Llandoverian flooding
sea

56 Geographical change in the lower Silurian

Climate of the past may be reconstructed from clues in both the fossil and the rock records. It is a general rule that a warm climate will support a more varied animal and plant life than a cold climate, so that a sequence of rock layers containing successively fewer fossil species may indicate progressive cooling of the climate, though there are other possible explanations. For the more recent geological eras fossils may be compared with closely related living forms of known climatic ranges and the ancient climate deduced. This has been done using foraminifera in ocean clays, insects, marine and freshwater molluscs, land vertebrates, and the pollen of land plants. A completely different method, the measurement of oxygen isotopes in fossil shells, allows estimation of water temperatures in the past to within 2 degrees C. Growing shells take up two isotopes of oxygen from water in a proportion which depends on temperature; this proportion is measured in the fossils using a mass spectrometer and the ancient temperature found. In one recent analysis it was found that Jurassic belemnites from Kawhia, New Zealand, lived in water with a seasonal range of 15°C to 20°C. Sediments in the rock record that give clues to climate in the past are red beds and wind-blown sands which formed under desert conditions, tills and outwash deposits laid down from melting ice, and laterite and bauxite formed by weathering in a tropical climate.

Application of these methods shows that the world today is climatically more varied than it generally has been in the past; in the Jurassic Period for example, there was probably little change in climate from the equator to the poles. On the other hand, the climate of any one place is bound to have changed with the passage of time. Figures 58, 59, and 60, from paintings in the Geological Museum, London, reconstruct the changing climate and conditions that can be deduced from strata in the Eden Valley in the north of England.

Changes like these may be explained in terms of the drift of continents through lines of latitude. But how can changes in the overall climate of the globe such as the great cooling that began 35 million years ago and culminated in the Pleistocene Ice Age, be accounted for? Internal factors which might affect the climate of the globe are the position of the continents, the distribution of land, sea and mountain ranges, and the amount of volcanic dust and carbon dioxide in the upper atmosphere. External factors are the variations in the orbit of the Earth, changes in the gravitational pull on the Earth by Sun, Moon and planets, and changes in the energy output of the Sun. The relative importance of all these factors is still in dispute, but it is clear that each ice age and each desert age has its own unique set of causes. The first two factors, position of the continents and the distribution of land and sea, are probably the most important, controlling the ocean currents and therefore the fall of snow and rain. But external factors such as the form of the Earth's orbit do certainly play a part, and their effect can be traced in the rapid climatic changes during the Pleistocene.

60 Eden Valley, Cumberland: warm, swampy forest 290 million years ago

59 Eden Valley, Cumberland: hot desert 250 million years ago

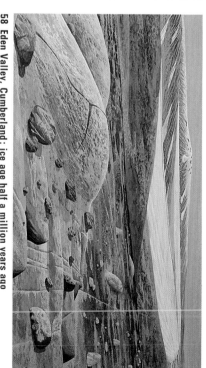

58 Eden Valley, Cumberland: ice age half a million years ago

Lower Palaeozoic geography. Positions of the three continents in existence 500 million years ago have been plotted using palaeomagnetic data. Evidence for large areas, such as China and southern South America, is very sparse, so there may be serious errors (fig 61). Very detailed mapping of the unconformity below the Silurian in the Welsh Borders of Great Britain has revealed the shape of the landscape which the sea submerged as it moved eastwards across central England 435 million years ago (fig 62). Some indications of surface geology at the time can also be given.

Carboniferous geography. Positions of the continents 325 million years ago can be plotted with rather more certainty than for the earlier period, though it is still only the latitude and alignment of fragments that are known; longitude and therefore the shape of the ocean basins is conjectural in both these Palaeozoic maps (fig 63). Studies of the different communities of marine animals that lived in the seas which covered northern England 310 million years ago have allowed a map to be drawn showing not only the ancient coastline, but also areas of particular depth and sea-floor conditions (fig 64).

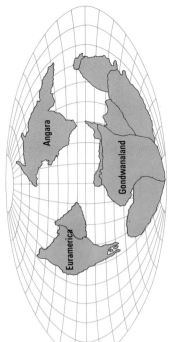

61 The continents 500 million years ago

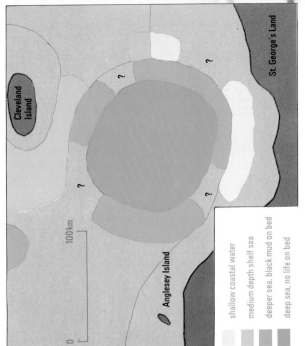

63 The continents 325 million years ago

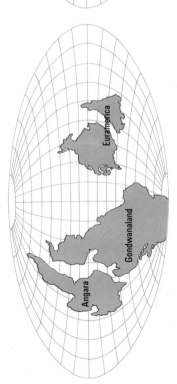

62 Welsh Borders 435 million years ago

O Ordovician
C Cambrian
P Precambrian

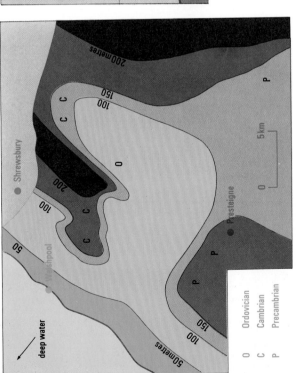

shallow coastal water

medium depth shelf sea

deeper sea, black mud on bed

deep sea, no life on bed

64 Northern England 310 million years ago

28

Jurassic geography. About 275 million years ago the three continents that had existed throughout the Palaeozoic collided and fused into a single mass, *Pangaea*. The Ural Mountains mark the line of collision of two of the ancient continents. This mass survived unchanged through the Permian and Triassic periods into the Jurassic, 175 million years ago (fig.65). The detailed map shows three islands of Carboniferous limestone that stood above the early Jurassic sea in South Wales (fig 66). A number of caves, filled with Jurassic sediment when the islands were submerged, can be traced in the limestone.

Eocene geography. The break-up of Pangaea, which started about 100 million years ago, led directly to the pattern of continents we know today. As early as the Eocene Epoch, 50 million years ago, the Atlantic, Indian, and Pacific oceans were in existence (fig.67). The collision of India with Asia and the separation of Australia from Antarctica are the major geographic events of the last 50 million years. The detailed map shows the geography of the English Channel at the time when the London Clay and the sands of the Western Province of the Channel were being deposited (fig.68).

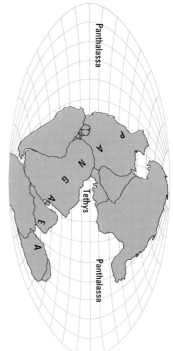

65 The continents 175 million years ago

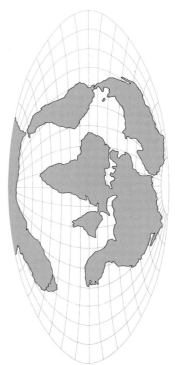

67 The continents 50 million years ago

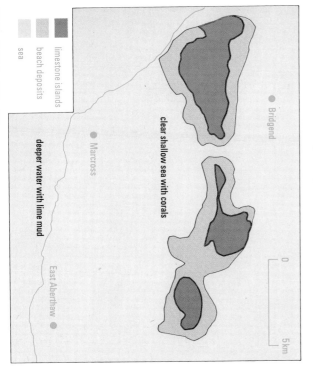

66 South Wales 175 million years ago

- sea
- beach deposits
- limestone islands

Bridgend

Marcross

East Aberthaw

clear shallow sea with corals

deeper water with lime mud

0   5km

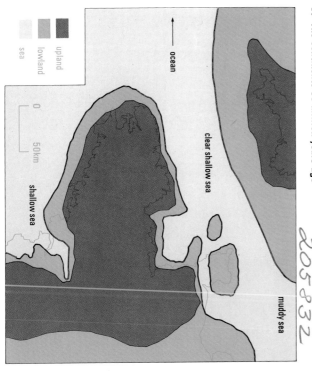

68 English Channel 50 million years ago

- sea
- lowland
- upland

ocean

clear shallow sea

shallow sea

muddy sea

0   50km

29

Life in the Palaeozoic

Life had a long history in the Precambrian which has left little record in the rocks. At the start of the Palaeozoic a number of unrelated animal groups, including sponges, archaeocyathids, brachiopods and trilobites, developed hard skeletons. They all lived in the sea and were mostly small. Throughout the era these groups, and others which appeared in the fossil record later on, evolved by natural selection to occupy new environments until, by the end, life was to be found in great variety in the air, on land, in freshwater, and in the sea. At the end of the Palaeozoic many marine groups and land plants became extinct.

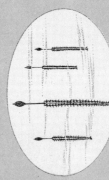

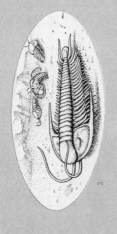

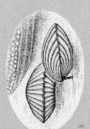

570 million years ago

500

1 sea snail, *Proplina*, 1 cm long
2 sponge-like archaeocyathids, 10 cm tall
3 trilobite, *Paradoxides*, 20 cm long
4 graptolite, *Climacograptus*, 5 cm long
5 brachiopod, *Onniella*, 0.5 cm across
6 giant water scorpion, *Pterygotus*, 2 metres long
7 nautiloid mollusc, *Orthoceras*, 50 cm long
8 sea lily, *Eycalyptocrinites*, 30 cm tall

30

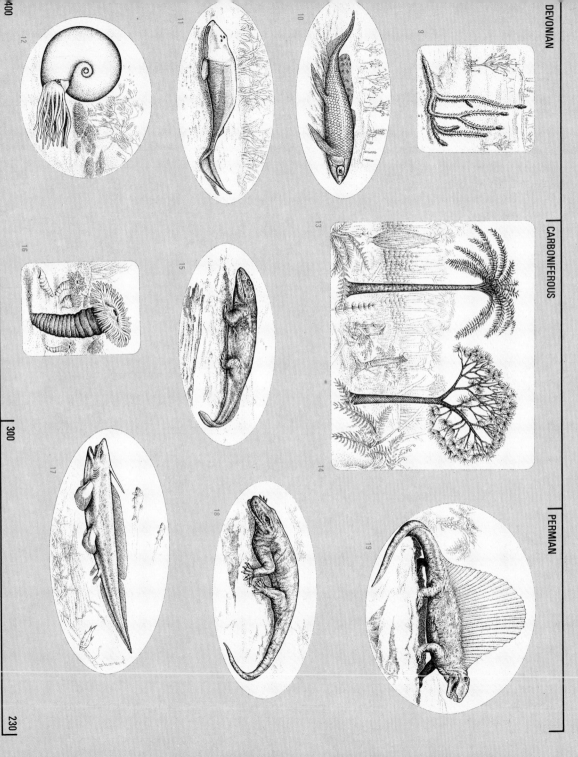

400

300

230

9 early land plant, *Protolepidodendron*, 30 cm tall
10 lungfish *Scaumenacia*, 30 cm long
11 primitive jawless fish, *Cephalaspis*, 25 cm long
12 goniatite mollusc, *Manticoceras*, 10 cm across

13 a seed-bearing tree fern, 15 metres tall
14 a spore-bearing tree fern, 20 metres tall
15 an early amphibian, *Ichthyostega*, 30 cm long
16 a rugose coral, *Palaeosmilia*, 10 cm long

17 an early shark, *Pleuracanthus*, 1 metre long
18 one of the first reptiles, *Captorhinus*, 1 metre long
19 a carnivorous reptile, *Dimetrodon*, 3 metres long

Life in the Mesozoic

TRIASSIC PERIOD

Late Palaeozoic extinctions left the early Triassic with a poor fauna and flora, and not until the Jurassic did evolution among the molluscs, corals and brachiopods make up for them. Ammonites and belemnites are the marine animals most characteristic of the era. On land, evolution of the late Palaeozoic reptile stocks gave rise to dinosaurs, birds, pterosaurs, crocodiles and primitive mammals, as well as the ichthyosaurs and plesiosaurs which returned to the sea. Ginkgos, bennetitales and sequoia trees flourished in the Jurassic, though flowering plants dominated the land flora by the end of the Cretaceous.

JURASSIC

230 million years ago

200

150

1 mammal-like reptile, *Cynognathus*, 3 metres long
2 ancestor of the dinosaurs, *Euparkeria*, 1 metre long
3 sea urchin, *Cidaris*, 1 cm across
4 group of corals, *Thecosmilia*, 10 cm across
5 extinct tree, a bennetitale, 3 metres tall
6 the first bird, *Archaeopteryx*, 50 cm wingspan
7 giant herbivorous dinosaur, *Brontosaurus*, 20 metres long
8 fish-like reptile, an ichthyosaur, 3 metres long

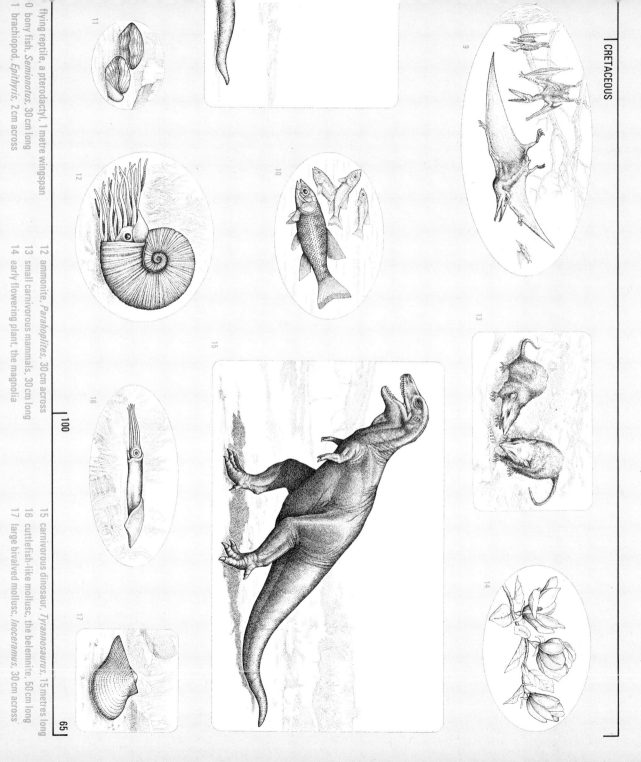

9

10

11

12

13

14

15

16

17

flying reptile, a pterodactyl. 1 metre wingspan

10 bony fish, *Semionotus*. 30 cm long

11 brachiopod, *Epithyris*. 2 cm across

12 ammonite, *Parahoplites*. 30 cm across

13 small carnivorous mammals. 30 cm long

14 early flowering plant, the magnolia

15 carnivorous dinosaur, *Tyrannosaurus*. 15 metres long

16 cuttlefish-like mollusc, the belemnite. 50 cm long

17 large bivalved mollusc, *Inoceramus*. 30 cm across

100

65

33

Life in the Tertiary

Ammonites, belemnites and most brachiopods having become extinct at the end of the Mesozoic, the Tertiary marine fauna was very similar to that of today. Most of the fish, molluscs and echinoderms of that time have close living relatives. Extinction of the dinosaurs was followed first by the evolution of large flightless birds and then by a spectacular radiation of the mammals into a group as varied and successful as the reptiles had been in the Mesozoic. Evolution of the flowering plants continued through the Tertiary to produce the great variety of trees, shrubs, vines and herbs that cover the land today.

65 million years ago   60   50   40

1 oppossum-like marsupial, *Thylacodon*, 30 cm long
2 oyster, *Ostrea*, 10 cm across
3 giant flightless bird, *Diatryma*, 2 metres tall
4 archaic carnivorous mammal, *Patriofelis*, 1 metre long
5 an early primate, *Smilodectes*, 50 cm tall
6 one of the first whales, *Basilosaurus*, 20 metres long
7 giant herbivorous mammal, *Brontotherium*, 3 metres tall

34

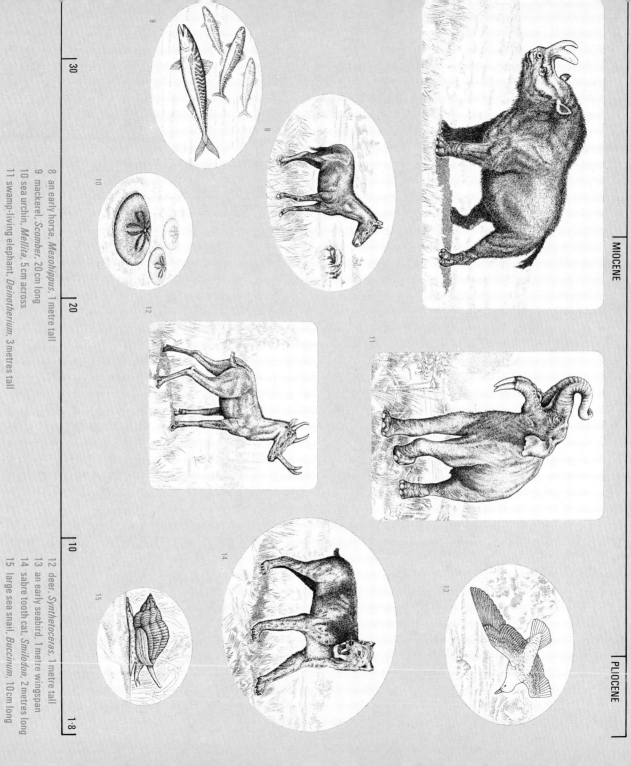

8 an early horse, *Mesohippus*, 1 metre tall
9 mackerel, *Scomber*, 20 cm long
10 sea urchin, *Mellita*, 5 cm across
11 swamp-living elephant, *Deinotherium*, 3 metres tall

12 deer, *Synthetoceras*, 1 metre tall
13 an early seabird, 1 metre wingspan
14 sabre tooth cat, *Smilodon*, 2 metres long
15 large sea snail, *Buccinum*, 10 cm long

30

20

10

1·8

The Quaternary was marked by a great ice age in the northern hemisphere associated with changes in sea level and the development of great deserts. It also saw the emergence and eventual dominance of Man. The differences between Pleistocene and modern living forms lie largely among the land mammals and are due to extinction rather than to evolution. Many large mammals have died out in the last hundred thousand years, including the mammoth and woolly rhinoceros of Europe, the giant kangaroo of Australia, and the enormous ground sloth from South America. So far there is not much sign of new forms developing to replenish the fauna. Man, having evolved in Africa during the Miocene, was a widespread and accomplished toolmaker by the beginning of the Pleistocene. *Homo sapiens*, modern man, evolved in Asia and moved into Europe 50000 years ago. Agriculture developed in the Middle East about 10000 years ago, and 4000 years later came the building of the first cities and the first use of writing. With these events we have reached the dawn of historical time and the end of geology.

From our place in the 20th century it is easy to exaggerate the special features of the world today, to overestimate the importance of Man on the Earth as a whole, and to feel that, with his arrival, geological time is somehow at an end.

The geologist, with the 4600 million years that is the age of the Earth always in mind, realises that the few thousand years of Man's dominance are only a drop in the ocean of geological time. He realises that, whether Man as a species survives for another hundred years or for a few million, time will go on, geological processes will go on, and life will go on. It may well be that Man, for all his commanding position today, will cut a very small figure in the geological record of the future.

Life in the Quaternary

PLEISTOCENE

| 1·8 million years | 1·0 | 0·5 |

1 giant Irish deer, *Megaceros*, 3 metres antlerspan
2 elephant of the Ice Age, the woolly mammoth, 3 metres tall
3 Neanderthal Man, lived in Europe 50000 years ago
4 South American ground sloth, *Megatherium*, 4 metres long